BLASTOFF!

NEPTUNE

BLASTOFF!

NEPTUNE

by Rebecca Stefoff

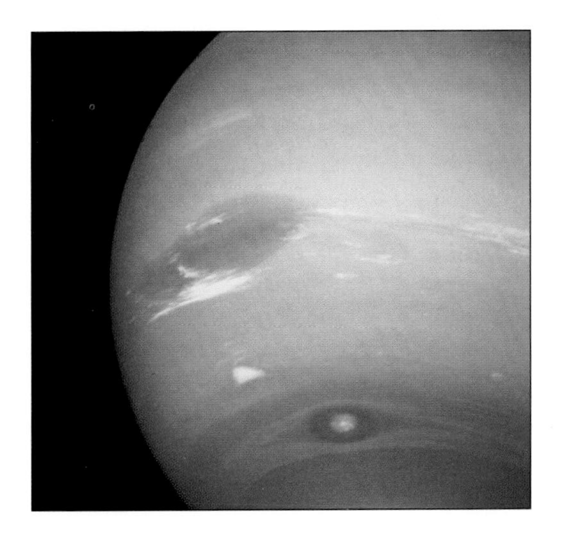

BENCHMARK BOOKS
MARSHALL CAVENDISH
NEW YORK

With special thanks to Roy A. Gallant, Southworth Planetarium, University of Southern Maine, for his careful review of the manuscript.

Benchmark Books
Marshall Cavendish Corporation
99 White Plains Road
Tarrytown, NY 10591-9001
www.marshallcavendish.com

Library of Congress Cataloging-in-Publication Data
Stefoff, Rebecca, date
Neptune / Rebecca Stefoff.
p. cm. —(Blastoff!)
Includes bibliographical references and index.
ISBN 0-7614-1232-8
1. Neptune (Planet)—Juvenile literature. [1. Neptune (Planet)] I. Title. II. Series.
QB691 .S74 2002 523.48'1—dc21 00-059643

Printed in Italy
1 3 5 6 4 2

Photo Research by Anne Burns Images
Cover Photo: Cosmographica/Don Dixon

The photographs in this book are used by permission and through the courtesy of: Corbis: Bettman, 9; Nasa/Roger Ressmeyer, 21. Sacro Bosco di Bomarzo, Lazio, Italy/Bridgeman Art Library: 14. NASA: 22. Photo Researchers: 8, 12, 13; NASA, 36; W. Kaufmann, 3, 16; Julian Baum, 11, 15, 24, 41; David A. Hardy, 19, 27; David Nunuk, 28; Jack Finch, 35; Seth Shostak, 55. JPL, 31, 37, 38, 42, 44, 49. Cosmographica/Don Dixon, 32, 56. M-Press Digital: 33, 45, 51, 52. ASP: Hartman, 46.

Book design by Clair Moritz-Magnesio

CONTENTS

1

THE OTHER BLUE PLANET

Ever since human beings first ventured into space in the 1960s, we have seen many pictures of Earth, our home planet, as it appears from orbit or from the Moon. Against a background of velvety blackness and glittering stars, the Earth is a deep, glorious blue. It is streaked with white clouds and patched with tan land masses, but most of it is azure as the water that covers two-thirds of its surface.

The Earth seems to be the only water-covered planet in our Solar System, but it is not the only blue one. Far out in the edge of the system rolls another blue marble, one much larger and colder than Earth. This is the planet Neptune. It is so far away that astronomers on Earth cannot study it in detail. Most of what we know about Neptune comes from the *Voyager 2* space probe, which passed close to the planet in 1989. In fact, Neptune's very existence was unknown through most of human history. It was discovered in 1846 by scientists who figured out where it was before they ever set eyes on it.

DISCOVERED ON PAPER

Neptune was the first planet whose existence was predicted on paper before the planet was identified through a telescope. To understand just how Neptune was found, we need to know a little about how planetary astronomy, the scientific study of the planets, developed over the centuries.

Cold and mysterious, the planet Neptune circles the Sun near the far fringes of the Solar System. Its discovery was one of the triumphs of nineteenth-century science.

Five of the planets in the Solar System are close enough, big enough, or bright enough to be easily seen from Earth with the naked eye. Night after night and month after month, they travel slowly along their courses among the fixed stars. Mercury, Venus, Mars, Jupiter, and Saturn have been known to people around the world for thousands of years. Ancient Babylonian astronomers, gazing into the clear night skies of the Middle East, first charted their movements. After the invention of the telescope in the early seventeenth century, scientists learned even more, discovering moons and rings around some planets and even spotting a few features on their surfaces. But as far as anyone knew or suspected, these five familiar sights were the only planets in the heavens.

Then, in 1781, a German musician living in England made a sensational discovery. William Herschel's hobby was astronomy, and he was passionately devoted to it. While making a detailed survey of the stars, Herschel found something that appeared to be a small round

William Herschel built this 40-foot (12.2-m) telescope in the 1780s at a cost of four thousand pounds—a substantial fortune at that time. A few years earlier, using another telescope, he had discovered the planet Uranus.

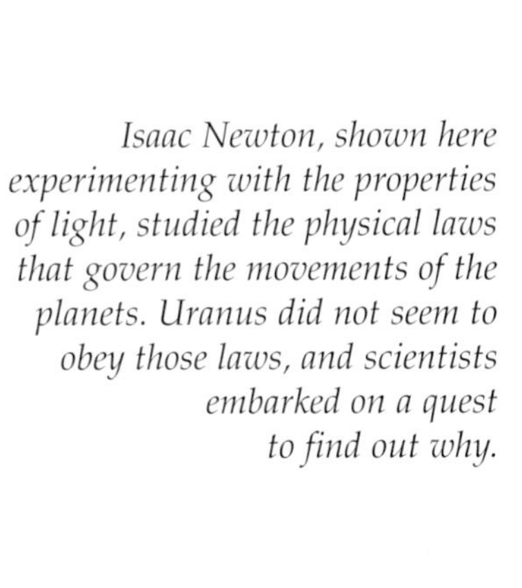

Isaac Newton, shown here experimenting with the properties of light, studied the physical laws that govern the movements of the planets. Uranus did not seem to obey those laws, and scientists embarked on a quest to find out why.

disk, not a point of light like the other stars. Surveying the new object over several nights, he saw that it moved across the starfield. Although at first he thought it was a comet, Herschel soon realized that this traveling body was a planet. Other astronomers agreed, and the new planet came to be called Uranus. For the first time, the family of planets had grown larger.

Uranus was the farthest known planet, revolving around the Sun in an orbit beyond Saturn's. Scientists were able to predict the path of Uranus's orbit because, at the end of the seventeenth century, an English mathematician named Isaac Newton had outlined the laws of motion, force, and gravity that govern all moving bodies, from billiard balls to planets. This system, called Newtonian physics, explained celestial motion, including the movement of the newly discovered planet.

But within a few decades of Uranus's discovery, astronomers realized that something unexpected was happening. Uranus was not obediently following the path that Newtonian physics predicted it would take. No one wanted to suggest that Newton's laws were inaccurate. The laws correctly explained how moving bodies move—everything except the new planet Uranus, which stubbornly continued to follow its strange new course. But if Newton's laws were correct, what could account for the unexpected motion of Uranus?

The puzzle perplexed Europe's astronomers, who came up with a number of clever explanations. Some believed that Uranus possessed a huge satellite. The gravitational pull of this satellite, they said, was strong enough to change Uranus's orbit. Others thought that Uranus had been struck by a comet and knocked out of its proper orbit just before Herschel discovered it. But most experts felt that none of these solutions could fully account for Uranus's movements.

At the beginning of the nineteenth century, a few scientists approached the right answer. They suspected that Uranus's orbit was influenced by another unknown body in the solar system. Newtonian physics said that such a body, if it were farther from the Sun than Uranus, would have a gravitational effect on Uranus, pulling the recently discovered planet out of the position it was "supposed" to occupy. Scientists called this perturbing the planet's orbit. By the 1830s, a handful of astronomers in Great Britain, France, and Germany suggested that the best explanation for the perturbations of Uranus's orbit was an undiscovered planet beyond Uranus, a planet so far away that, like Uranus, it was invisible to the naked eye.

A Scottish scientific writer named Mary Fairfax Somerville wrote in 1836 that Uranus "may be subject to disturbances from some unseen planet revolving about the sun beyond the boundaries of our present system." She went on to suggest that careful study of the perturbations of Uranus's orbit over a few years "may reveal the existence, nay, even the mass and orbit, of a body placed forever beyond

The sighting of Uranus, the seventh planet from the Sun, greatly enlarged our conception of the Solar System. It would be more than two hundred years before Voyager 2 *gave us a closer look at Uranus's complex world.*

the sphere of vision." Five years later a German astronomer named Johann Heinrich von Mädler wrote that the discovery of such a planet would be the "highest triumph" of mathematics, "a discovery made with the mind's eye, in regions where sight itself was unable to penetrate."

Uranus had been discovered by accident, during a telescopic survey of the stars. Now a few scientists were suggesting that another planet might await discovery. And complicated mathematical calculations—based on the difference between where Uranus was supposed to be and where it actually was—could tell the astronomers where to point their telescopes to find that mystery planet.

SEARCHING THE SKIES

Two young mathematicians, one in France and one in England, took up the challenge. In England, John Couch Adams wrote in his journal that as soon as he graduated from college he would devote himself to the problem of Uranus, with the hope of pinpointing the location of "an undiscovered planet" by analyzing its effects upon Uranus. Adams graduated from Cambridge University in 1843 and set to work.

Two years later Adams felt that he knew the general location of the mystery planet. He tried several times to get Sir George Airy, the astronomer royal of Great Britain, interested in his findings, but even with the help of a Cambridge astronomy professor Adams could not awaken any enthusiasm in Airy. Finally, Adams wrote Airy a letter containing the results of his calculations. He thought that Airy would start looking for the mystery planet at once, using the powerful telescopes of the Royal Observatory at Greenwich, outside London. Instead, Airy wrote back to Adams asking a question about the calculations. When Adams read Airy's letter, he was convinced that the astronomer royal did not understand his solution to the problem.

On the other side of the English Channel, in France, a mathematician named Urbain-Jean-Joseph Leverrier was also investigating the

John Couch Adams was one of Neptune's discoverers, although he did not receive credit for his work until others had claimed the find. Born the son of a tenant farmer, Adams received his education on a scholarship and later refused the offer of a knighthood.

The calculations of French science teacher and mathematician Urbain-Jean-Joseph Leverrier led to the first sighting of the "mystery planet"—and prompted an international scientific disagreement.

perturbations of Uranus, working independently of Adams. By the end of 1845 Leverrier had published his first set of results, showing that an undiscovered planet could produce the Uranian perturbations. By mid-1846 he had published a complete solution that placed the unknown planet exactly where Adams had said it was. Airy, the English astronomer, wrote to Leverrier, congratulating him on his work, but he did not mention Adams or tell Adams of Leverrier's solution.

Airy considered himself and the Greenwich Observatory too busy to look for the mystery planet, so he asked the Cambridge astronomy professor to do so. With many delays and setbacks, the British search began. Meanwhile, Leverrier was impatient with the Paris Observatory. He wanted an immediate search for his planet, but the director of the observatory was slow in starting one as well—despite the fact that he had asked Leverrier to locate the planet in the first place.

In September 1846 Leverrier sent his calculations to the Berlin Observatory, asking the astronomers there to look for his planet at once. The very night they received his letter, astronomer Johann Gottfried Galle and a student named Heinrich D'Arrest turned a telescope to the spot where the planet was predicted to be. They began checking each point of light against a star map. After they had been working for just a few minutes, Galle, who was looking through the telescope, called out the position of another glowing body. "That star is not on the map!" cried D'Arrest. The mystery planet had been found.

The Berlin astronomers tracked the new planet's course for the rest of the night and then observed it the following evening. It appeared exactly where Leverrier's calculations had said it would be. The director of the Berlin Observatory wrote to the French mathematician, congratulating him on his "brilliant discovery . . . all that the ambition of a scientist can wish for." Finally scientists had bridged the millions of miles separating Earth and Neptune, as the new planet was called.

The Sea God in Space

Mercury, Venus, Mars, Jupiter, Saturn, and Uranus are named for gods and goddesses of the ancient Romans. When people began looking for an unknown planet beyond Uranus, most agreed that it, too, should receive a name from Roman mythology. Leverrier was one of many who suggested Neptune, after the Roman god of the sea. Once the planet had been discovered, however, he tried to convince everyone that it should be named Leverrier. Finally, though, even he admitted that Neptune was more suitable. The sea god's name, in fact, was a very good choice for a sea-blue planet.

Of course, the glorious news from the Berlin Observatory was met with dismay in Great Britain. The astronomer in Cambridge had been steadily working his way through a section of the sky, and he had even spotted something that he thought might be a planet. But before he could double-check this find, Galle's discovery—linked with Leverrier's name—had been announced.

British scientists were now eager to claim a share of the glory for their nation. Sir John Herschel, son of the discoverer of Uranus, pointed out in the press that although Leverrier's calculations had certainly led to Neptune's discovery, a British mathematician, Adams, was actually the first person to solve the mystery of the planet's location. Sir George Airy, whose failure to listen to Adams had probably cost him the chance to discover Neptune, tried to smooth things over by

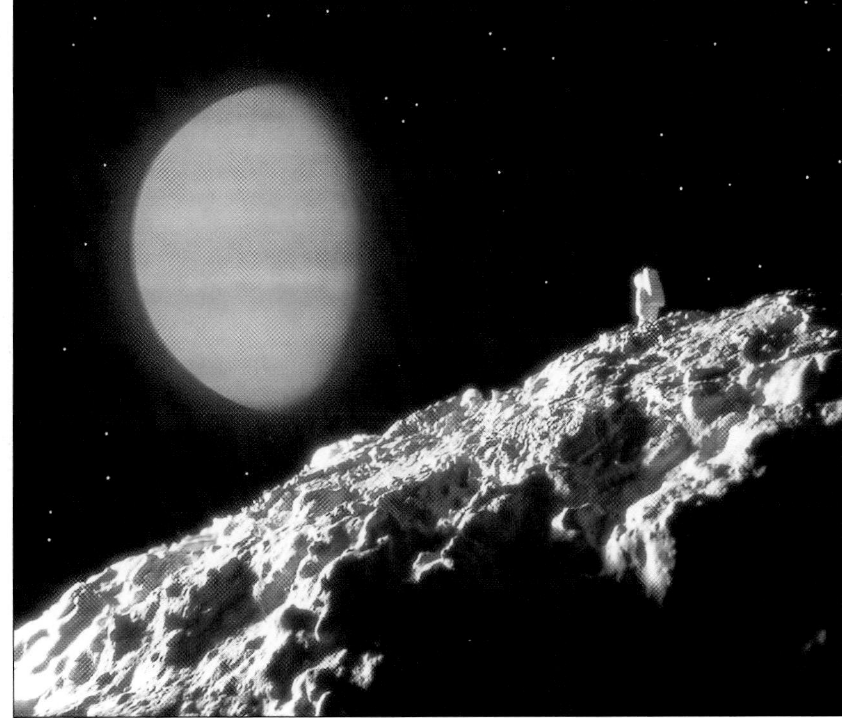

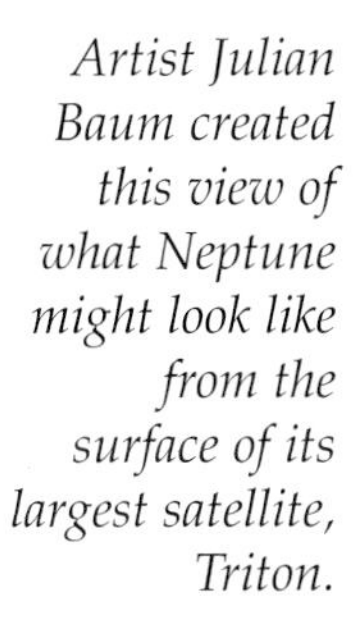

Artist Julian Baum created this view of what Neptune might look like from the surface of its largest satellite, Triton.

Just in Time?

After Neptune had been discovered, astronomers tracked its orbit and took a second look at the calculations of Adams and Leverrier. Surprisingly, they found that both had based their work on incorrect predictions of Neptune's distance from the Sun. One astronomer claimed that Leverrier's calculations were so far off that Galle's sighting of the planet was "a happy accident." Most modern experts, however, feel that while Adams's and Leverrier's predicted orbits were somewhat inaccurate, both men were close to the correct solution. Still, everyone involved in the discovery of Neptune was lucky that it happened when it did. Between 1800 and 1850 Neptune's real orbit closely matched the path predicted by Adams and Leverrier. If astronomers had used the same calculations to search for the planet just a few years later, Neptune might not have been found so easily.

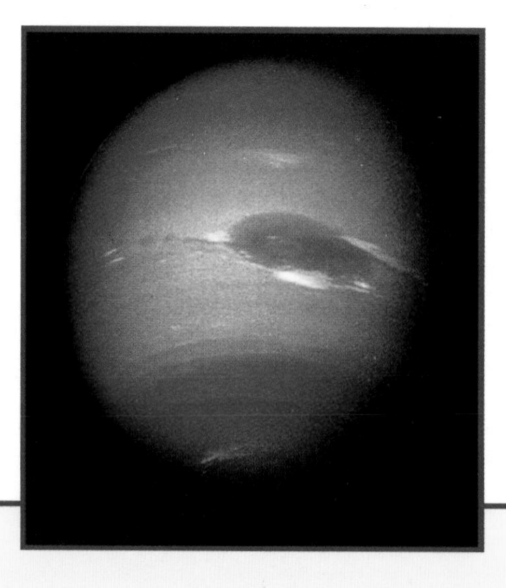

claiming that Adams had solved the problem first but that Leverrier's answer was more complete.

Once they realized that the British were staking a claim to the discovery of Neptune, French scientists responded with outrage. (France and Great Britain, who had often been at war during the preceding century, were great rivals in science and many other fields.) One Frenchman wrote, "Mr. Adams has no right to figure in the history of the planet, neither by a detailed citation, nor by the slightest allusion." The Neptune controversy raised tempers on both sides of the English Channel and almost became an international incident.

Eventually, however, the general public and the international scientific community calmed down. After a great many letters and statements about the matter had been published, most people agreed that Adams and Leverrier had independently calculated Neptune's position at almost exactly the same time and that Galle and D'Arrest had been the first to identify the planet. Both Adams and Leverrier received credit for performing the essential calculations that located the planet. In addition, both of them deserve credit for not getting involved in the argument over national glory. Each man behaved in a respectable manner, and when they met for the first time in 1847, they started a lifelong friendship.

Once Neptune's orbit became known, astronomers combed old records to see if any earlier skywatchers had spotted Neptune without knowing that it was a planet. During the eighteenth and early nineteenth centuries, several astronomers had recorded sightings of Neptune but had marked it on their charts as a star. Even more amazing, the Italian scientist Galileo Galilei, one of the first people to turn a telescope toward the heavens, seems to have spotted Neptune several times in 1612 and 1613, while he was mapping the sky near Jupiter. He, too, thought that the bright object was a star. Not until scientists possessed the twin clues of Uranus's orbit and Newton's laws would they discover Neptune, the second planet found in modern times.

2

Flight of a Voyager

In the late twentieth century, the *Voyager 2* spacecraft gave scientists their best views of Neptune. Even before that time, however, astronomers had unlocked a few of the cloud-wrapped planet's secrets.

THE EIGHTH PLANET—USUALLY

The discoverers of Neptune thought that it was the eighth planet from the Sun, after Mercury, Venus, Earth, Mars, Jupiter, Saturn, and Uranus. It is—most of the time. Sometimes, however, it is the ninth planet. Neptune does not really change position. Its neighbor, Pluto, does. Pluto is the outermost known planet of the Solar System, discovered in 1930. Its unusual orbit passes considerably closer to the Sun at one end than at the other. Because of this, Pluto's path periodically crosses that of Neptune.

Neptune travels around the Sun once every 165 years—this is the planet's period of revolution, or one Neptunian year. Pluto's period of revolution is much longer—248 years. During most of its revolution, Pluto is the ninth planet. But for twenty years of every revolution, Pluto travels inside Neptune's orbit, making it the eighth planet and Neptune the ninth. On January 21, 1979, Pluto crossed inside the orbit of Neptune, and it stayed there until May 14, 1999. During that time, Neptune was the outermost planet of the Solar System. It will not be so again until the twenty-third century.

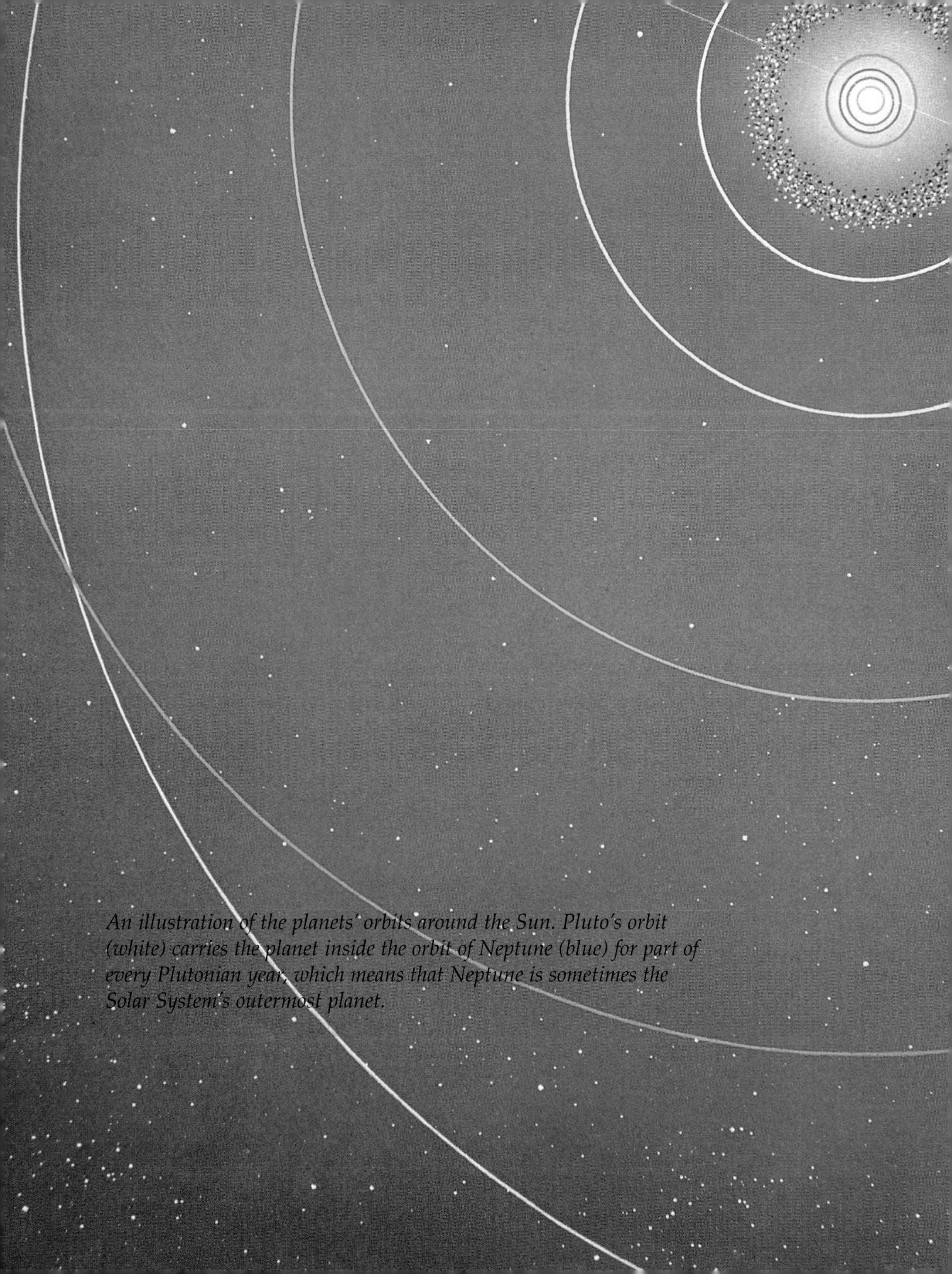

An illustration of the planets' orbits around the Sun. Pluto's orbit (white) carries the planet inside the orbit of Neptune (blue) for part of every Plutonian year, which means that Neptune is sometimes the Solar System's outermost planet.

Measuring Distance in Space

Miles and kilometers are useful for measuring distances on Earth. But the distances between most objects in space are so vast that scientists have developed units of measurement especially for them. Two of the units often used for measuring distance in space are the light-year and the astronomical unit, or AU.

A light-year is the distance that light travels in one year. Traveling at a constant speed of 186,000 miles per second (299,338 km/sec), light covers about 5,878,000,000,000 miles, or almost 6 trillion miles (9.7 trillion km), in a year. The light-year is usually used to measure distances outside the Solar System.

An astronomical unit is the mean distance from the Earth to the Sun, about 93 million miles (150 million km). The mean distance is the midpoint between the greatest distance and the least distance. Astronomers usually use AUs when comparing distances within the Solar System. The Earth, for example, is 1 AU from the Sun, while Venus is 0.72 AU from the Sun.

Here is Neptune's mean distance from the Sun, in a variety of units:

2,797,000,000 miles (almost 2.8 billion)
4,497,000,000 kilometers (almost 4.5 billion)
30.1 AU (more than 30 times as far as Earth)
0.000476 light-years (almost 5 ten-thousandths of a light-year—compared with a distance of 4.2 light-years from the Sun to the nearest star, Alpha Centauri)

FIRST VIEWS OF NEPTUNE

Great Britain had missed the chance to discover Neptune, but one British astronomer was determined not to miss the next sensational find. As soon as Sir William Lassell learned of Neptune's existence and location, he began studying the planet through his own large telescope. He was especially curious to learn whether Neptune had any satellites, or moons, like those known to revolve around Jupiter, Saturn, and Uranus.

A twentieth-century photograph of Triton. William Lassell's view of the moon looked nothing like this—nineteenth-century astronomers would have seen only a tiny blob of light in Neptune's vicinity, or a dark speck silhouetted against its surface.

In October 1846 Lassell found a Neptunian moon, Triton. Lassell also believed that he had seen rings around Neptune, like the rings that encircle the planet Saturn. A few other astronomers also claimed to have seen Neptunian rings, but the sightings were dim and unclear, and many who looked for rings failed to find any trace of them. After a few years the idea of Neptunian rings was abandoned.

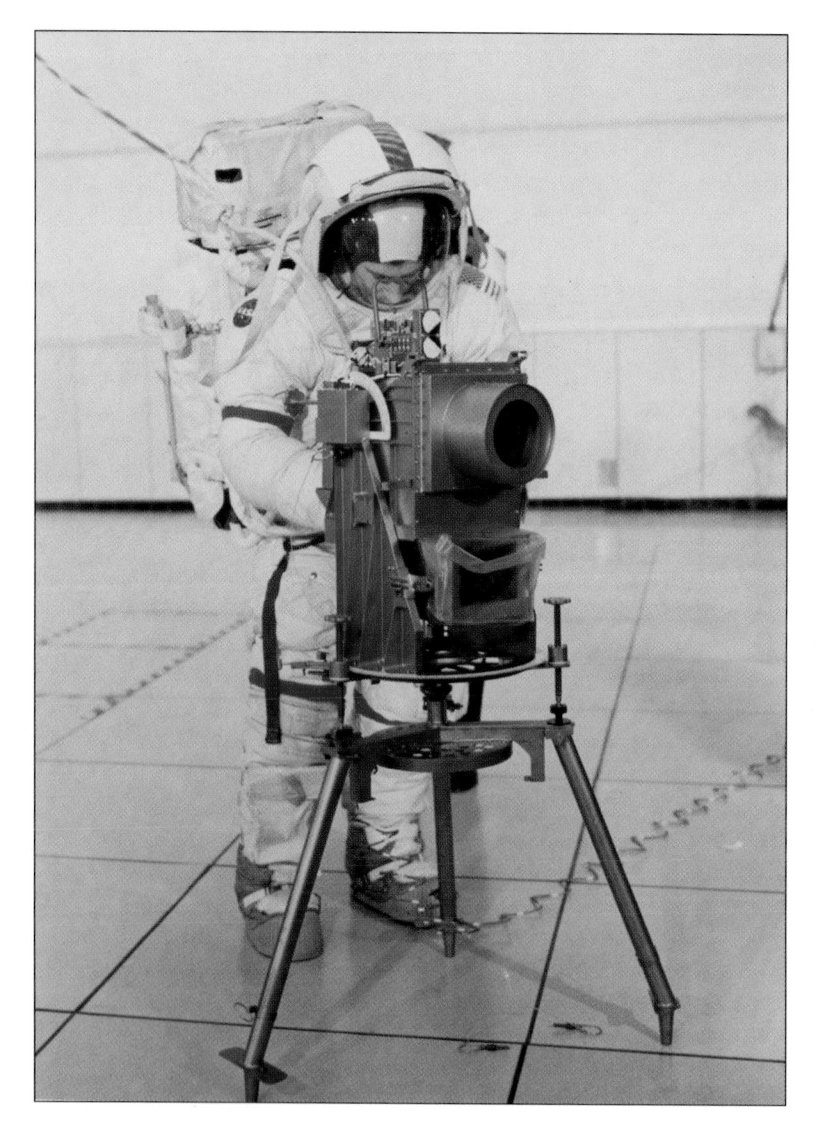

Astronaut John W. Young, shown here in 1971, is trained in the use of a spectroscope that was later placed on the Moon's surface. Similar instruments gave Earth-based scientists their first information about Neptune's chemical makeup.

Astronomer Gerard Kuiper discovered Neptune's second moon, Nereid, in 1949. He found it not by gazing at the sky but by studying photographs taken through a powerful telescope in Texas. Comparing a series of images of the same part of the sky, Kuiper noticed one body that moved over time. Soon he had identified it as a new Neptunian satellite.

One lingering mystery concerned Neptune's period of rotation, or the length of the planet's day. Astronomers had determined the periods of rotation of Jupiter and Saturn by noting features on the planets' surfaces and measuring how long those features took to return to the same positions as the planets rotated. But Uranus and Neptune are so far away and so featureless that even the sharpest-eyed astronomers, using the biggest telescopes, had a hard time finding any identifying marks on their surfaces.

By the 1980s, scientists had used instruments called spectroscopes to analyze the light coming from Neptune. Spectroscopes break light into broad bands. Vertical lines within these bands form patterns that indicate the chemical elements making up the light source. By comparing spectroscopic patterns from Neptune with those of known

Can You See Neptune?

Neptune is too far away for anyone to see it with the naked eye. But if you knew exactly where to look, you could see Neptune through a good pair of binoculars. It would look like a faint star. In order to see Neptune as a tiny glowing disk instead of a point of light, you would need a telescope with an aperture, or opening, at least 5 inches (12.7 centimeters) across.

Julian Baum's vision of Voyager 2 *leaving the vicinity of Triton, the last body the spacecraft visited in our Solar System.*

chemicals, scientists could identify the elements on Neptune that were reflecting sunlight. They found that the planet is covered with gases, mostly hydrogen. In their telescopic surveys, they had spotted what seemed to be huge clouds in the Neptunian atmosphere, and they had seen faint hints of the rings Lassell claimed to have seen more than a century earlier. They estimated Neptune's period of rotation was about 18 hours, based on the faint and uncertain sightings of these cloud features. But there was a limit to their discoveries. Earth-based astronomers could learn little more about Neptune until they could somehow get closer to the planet.

A HISTORIC FLYBY

Astronomers got a close-up look at Neptune in 1989. No one traveled there in person—such a far-ranging manned spaceflight is still many years in the future. But an unmanned spacecraft called *Voyager 2* flew close to Neptune on its way out of the Solar System.

Neptune was the last stop in what space scientists called the Grand Tour of the outer planets. The U.S. National Aeronautics and Space Administration (NASA), which directed the *Voyager* program, took advantage of an event that happens only once every few hundred years in the Solar System. During the late 1970s and the 1980s, the planets Jupiter, Saturn, Uranus, and Neptune were all positioned in the same general direction from the Sun. This rare alignment allowed *Voyager 2* to visit each of these large, slow-moving planets in turn. Using cameras and other instruments controlled by an onboard computer system, *Voyager 2* observed Neptune from June to October 1989.

On August 25, the spacecraft made its closest approach to Neptune, passing just 3,000 miles (4,850 km) above the surface of the clouds over the planet's north pole. Five hours later, it traveled past

Missions to Deep Space

Voyager 2 was not even supposed to go to Neptune. Congress originally authorized NASA to send probes to examine Jupiter and Saturn, the largest and closest gas giants. In 1977 two nuclear-powered spacecraft, *Voyager 1* and *Voyager 2*, were launched. *Voyager 1* flew by Jupiter in 1979 and Saturn in 1980. It kept going, heading out of the Solar System into deep space, but its flight path did not take it near any other planets.

Voyager 2 passed Jupiter in 1979 and Saturn in 1981. Scientists realized that they could use Saturn's gravitational force to send the spacecraft on to Uranus and even perhaps to Neptune. Congress approved extending *Voyager 2*'s mission, and although the spacecraft had been designed to function for five years, it continued performing well long after that time. The 1,819-pound (825-kg) *Voyager 2* encountered Uranus in 1986 and Neptune three years later. Someday, perhaps, one or both of the *Voyager* spacecraft, which carry messages from Earth in the form of pictures and sound recordings, will be found by distant beings among the stars.

This illustration shows Voyager 2 *flying around Jupiter, using the slingshot effect of the giant planet's gravity to propel itself onward. Later the spacecraft employed a similar maneuver around Saturn to gain momentum for its journey to Uranus and Neptune.*

Triton, Neptune's moon, at a distance of 25,000 miles (40,000 km). By that time, the probe had been traveling for 12 years at an average speed of almost 42,000 miles an hour (67,500 km/hr) since being launched from Cape Canaveral, Florida.

The computer system aboard *Voyager 2* turned the information and images gathered by the spacecraft's cameras and sensors into radio signals that were sent back to Earth. There, scientists decoded

The VLA, or Very Large Array, antennas near Socorro, New Mexico, are part of the world's largest radio telescope. These and other "dishes" all over the world listened for the faint whispers of Voyager 2*'s transmissions from the distant reaches of the Solar System.*

the signals, turning them into pictures and other forms of data. The scientists could also radio commands to the spacecraft.

Communication was not easy, though. Neptune is so far away from Earth that by the time *Voyager 2* approached the planet, radio signals took more than four hours to travel between the spacecraft and Earth. The signals coming from *Voyager 2* were so faint that scientists had to use the world's largest antennae to receive them. Research stations in California, New Mexico, Spain, Australia, and Japan devoted antennae to the effort.

The result was greater than *Voyager's* creators had dared to hope. The spacecraft sent thousands of images and millions of pieces of other data to Earth. Almost as soon as the first images were received, scientists knew far more about Neptune than they had learned in the century and a half since the planet's discovery. And in the years that followed the flyby, as experts studied the data on Neptune and its surroundings, they gained hundreds of new insights—not just about Neptune but about how the entire Solar System is organized, perhaps even about how it was formed.

With its visit to Neptune complete, what happened to *Voyager 2*? It kept traveling beyond Neptune and is now headed toward the heliopause, the remote boundary between our Solar System and interstellar space. As the spacecraft completed its close encounter with the fourth gas giant, one member of the *Voyager* science team exclaimed, "What a way to leave the Solar System!"

3

A Gas Giant

Using data from the *Voyager 2* flyby, scientists have put together a new and much more complete picture of Neptune. Observations from the Hubble Space Telescope, which views the heavens more clearly than any Earth-based telescope because it is above the Earth's atmosphere, are adding to that picture. Neptune, it turns out, is a fascinating place.

ATMOSPHERE AND STRUCTURE

The Solar System has two kinds of planets. Mercury, Venus, Earth, and Mars, sometimes called the inner planets, are solid, rocky worlds. Pluto is either rocky or icy—either way, it is a small, hard lump of a planet. But Jupiter, Saturn, Uranus, and Neptune are entirely different. They are much farther from the Sun than the inner worlds. They are also much larger than any of the other planets, and neither solid nor rocky. Their surfaces are covered with thick layers of gases and liquids, although the planets may have solid or slushy cores deep within those layers. Together, these four planets are known as the gas giants.

Neptune is the fourth-largest planet in the Solar System and the smallest of the gas giants. The planet measures 31,404 miles (50,538 km) across at its equator. This equatorial diameter is slightly less than that of Uranus, but four times greater than Earth's. Neptune's polar diameter, the distance from its north to its south poles, is 30,821 miles (49,600 km). The polar diameter is a bit less than the equatorial

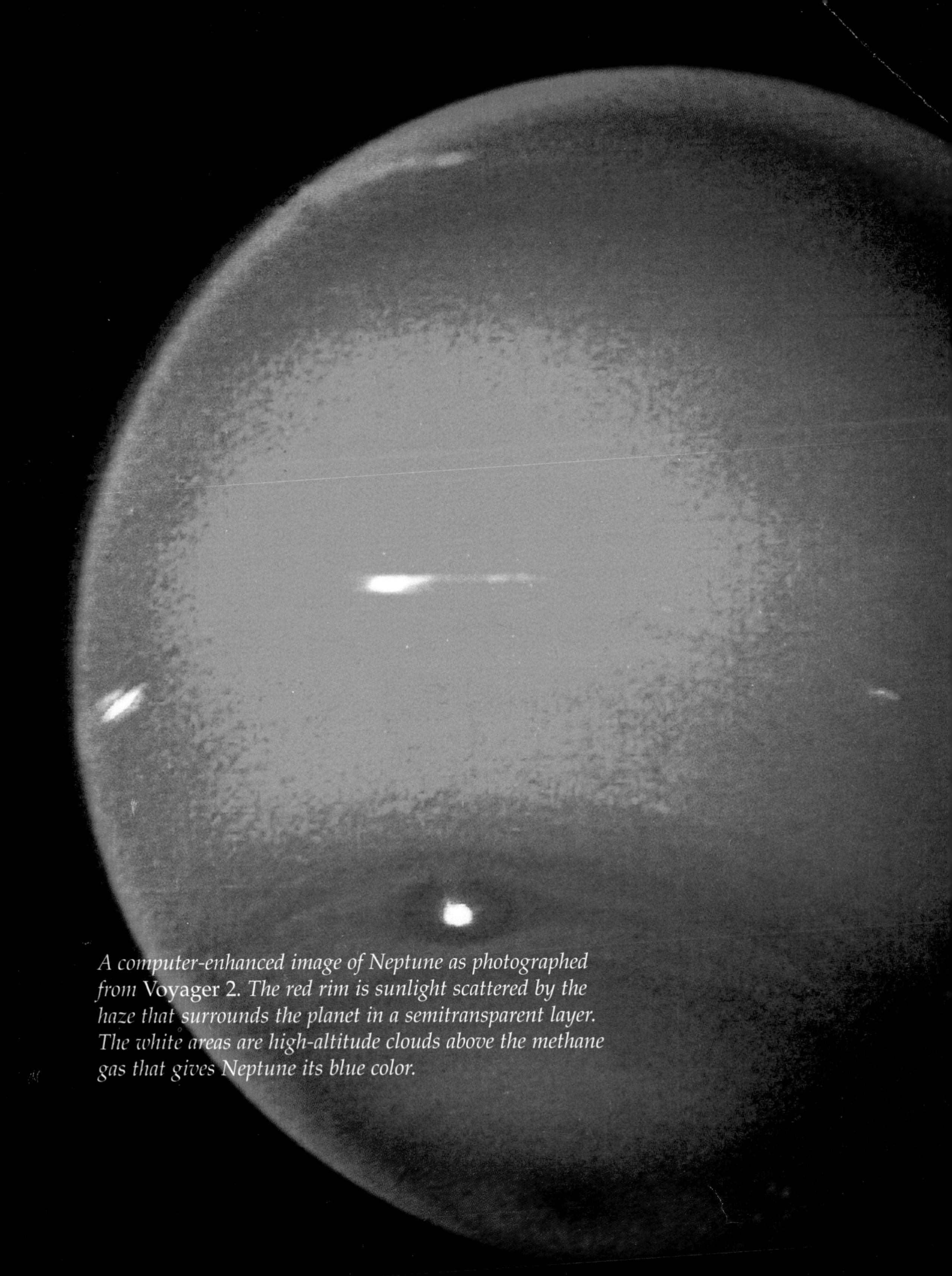

A computer-enhanced image of Neptune as photographed from Voyager 2. *The red rim is sunlight scattered by the haze that surrounds the planet in a semitransparent layer. The white areas are high-altitude clouds above the methane gas that gives Neptune its blue color.*

diameter, which means that Neptune is bigger through the middle than it is from top to bottom, like an orange that has been slightly flattened. Its volume is fifty-seven times that of Earth. If Neptune were hollow, you could squeeze fifty-seven Earths inside it—if you could pack them together tightly enough.

The only part of Neptune that can be viewed directly is the top layers of its cloud-filled atmosphere. That atmosphere would be deadly to a human being. Between 77 and 83 percent of it is hydrogen gas. Another 16 to 20 percent is helium. Methane accounts for 1 to 2 percent, and the rest of the atmosphere consists of small amounts of ethane, ammonia, tiny suspended particles of water ice, carbon monoxide, and hydrogen cyanide. It is the methane in Neptune's atmosphere that gives the planet its bright blue color. This gas absorbs the red wavelengths in the sunlight that reaches the planet, so that what is reflected from the cloud cover is primarily blue.

No one is certain just what Neptune is like below its dense covering of cloud, but scientists have developed two main ideas about the planet's structure. Some believe that under the gas layers Neptune

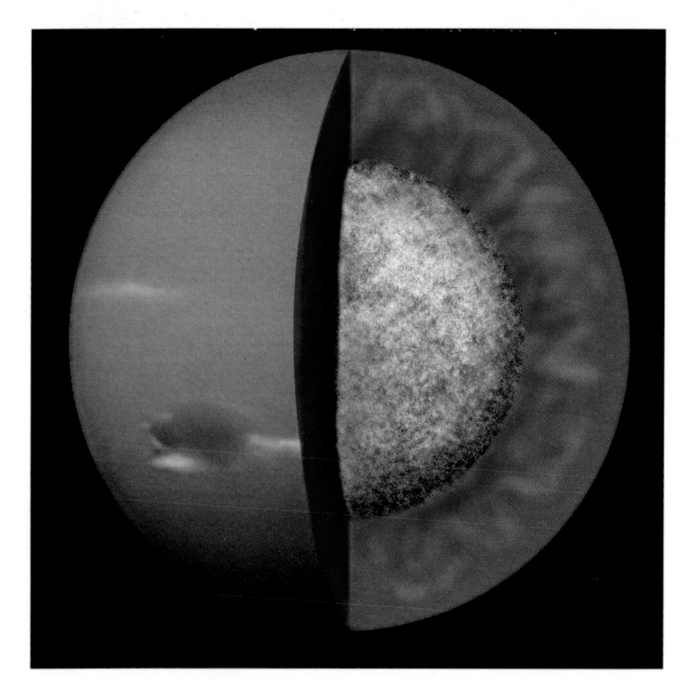

Neptune's interior structure remains unknown. Some scientists believe that a hard rocky core lurks beneath layers of gas and liquid. Others speculate that the core consists of pressurized liquids and melted rock.

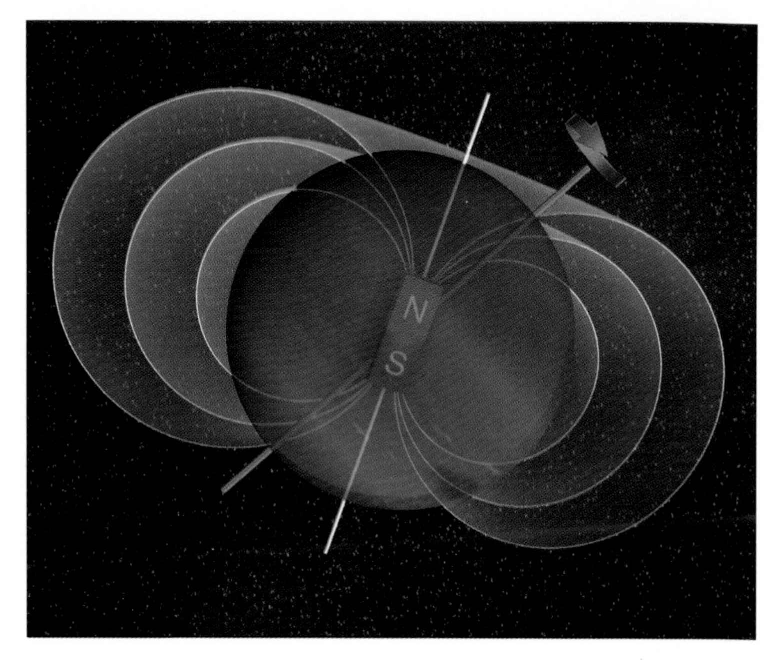

A magnetic mystery—Neptune's magnetic poles (blue) are tilted sharply away from its geographic poles (red), the points around which the planet rotates. Neptune's unusual magnetic features are probably caused by electrical activity deep within the planet.

consists of a giant ocean made up of water or other liquids. Within this there may be a rocky core about the size of Earth. Others think that beneath the cloud cover is a shell about 6,000 miles (10,000 km) thick, made up of gas and slushy, partly melted ices. Deep within that shell, in a region of very high temperature and pressure, is a core of liquids and melted rock. Either way, Neptune seems likely to have a fairly uniform internal composition, unlike Jupiter and Saturn, which scientists think consists of many different layers.

MAGNETISM AND TEMPERATURE

As *Voyager 2* approached Neptune, scientists anticipated receiving the answers to some of their lingering questions. Researchers had wondered, for example, whether Neptune had a magnetic field. *Voyager 2* answered that question. Neptune possesses a strong and unusual magnetic field.

On Earth, the north and south magnetic poles are located fairly close to the north and south geographic poles, the points around which the Earth rotates. On Neptune, however, the magnetic poles are more than halfway to the planet's equator, so that Neptune's magnetic field is strongly tilted. In addition, it is greatly off-center. The field surrounds the planet like an invisible sphere, but its center is not located in the middle of Neptune. Instead, the center of the magnetic field is about 8,500 miles (13,500 km) from the center of the planet. This means that the electrical currents produced by magnetism within the planet are closer to its surface on one side of the planet than on the other. If you took a magnet to this part of Neptune, its attracting power would be strongest there.

The presence of a magnetic field offers hints about Neptune's structure. Scientists think that a magnetic field can exist around a world only if there is a liquid region within the planet, if that liquid is a material (such as water or melted metal) that conducts electricity, and if some energy source causes that liquid to move and to keep moving. The field of magnetic energy around Neptune may be caused by electrical activity within a layer or core of the planet that is continually sloshing or shifting about.

Neptune's magnetic field produces auroras, or elecromagnetic discharges in the atmosphere. Auroras are caused by electrically charged atomic particles called protons and electrons that the Sun discharges. As they enter a planet's magnetic field, these particles sometimes glow green, white, or other colors, producing rippling sheets or bands of light in the sky along the lines of the magnetic field. On Earth auroras are most often seen near the north and south magnetic poles. Neptune's auroras are fainter than Earth's, but they occur over wider areas of the planet, not just near its magnetic poles.

Voyager 2's measurements of Neptune's magnetic field helped scientists settle the question of the planet's period of rotation. As a planet rotates, its magnetic field gives off radio waves at regular intervals.

Earth's northern lights, or aurora borealis, as seen from Fairbanks, Alaska. Similar auroras illuminate Neptune's night skies.

These waves are a highly accurate guide to the amount of time the planet takes to spin completely around. *Voyager 2*'s instruments recorded the radio waves from Neptune's magnetic field. Analyzing this data, scientists discovered that although Neptune's year is 165 Earth years long, its day is shorter than an Earth day. Neptune rotates once every 16 hours and 7 minutes. However, the difference between day and night is far less noticeable on Neptune than it is on Earth. If you were on Neptune, you would not see the Sun as a glowing disk covering the world with light and warmth. At high noon in the middle of a Neptunian summer, the Sun would be little more than the brightest star in the black sky.

Even though it is far from the Sun, Neptune is not without heat. In fact, although it is extremely cold by Earth standards, Neptune gives off nearly three times as much heat as it absorbs from the Sun.

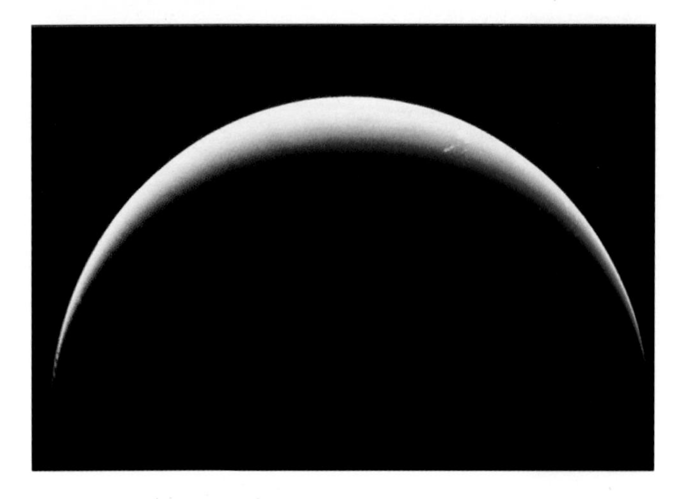

A Voyager 2 *image of Neptune's south pole. Neptune's poles, unlike Earth's, are about the same temperature as its equator—all are very cold, by Earth standards.*

Like Jupiter and Saturn, Neptune has a strong internal heat source (Uranus does not). Some of Neptune's inner heat may be left over from the formation of the Solar System. Some of it is probably created by intense pressure at the planet's core. In addition, scientists believe that Neptune's atmosphere generates heat through the movement of gases in a cycle known as convection. Cool gases sink. As they descend into the denser parts of the atmosphere, pressure heats them. This causes them to rise again, carrying heat up with them.

All of this activity, however, does not make Neptune balmy. *Voyager 2* measured the temperature near the bottom of the planet's

Wild Winds

Neptune is dark and cold, but it is far from quiet. Winds churn its atmosphere. In fact, scientists believe that Neptune is the windiest planet in the Solar System. Most of its winds have speeds of as much as 730 miles (1,175 km) an hour. The strongest winds measured, however, were blowing at 930 miles (1,500 km) an hour. On Earth, a wind blowing 74 miles (119 km) an hour is considered a hurricane!

atmosphere at -360 degrees F (-218 degrees C). On Earth, the hottest place is the equator and the coldest places are the poles, because the equator is closest to the Sun and the poles are farthest away. But *Voyager 2* found temperatures at Neptune's poles to be nearly the same as those at its equator. Scientists have not yet been able to explain why this is the case.

SPOTS AND A SCOOTER

Even before the *Voyager 2* flyby, astronomers had begun to make out a few surface features on Neptune. *Voyager 2* and the Hubble Space Telescope have given them much clearer views of these features in the planet's upper atmosphere.

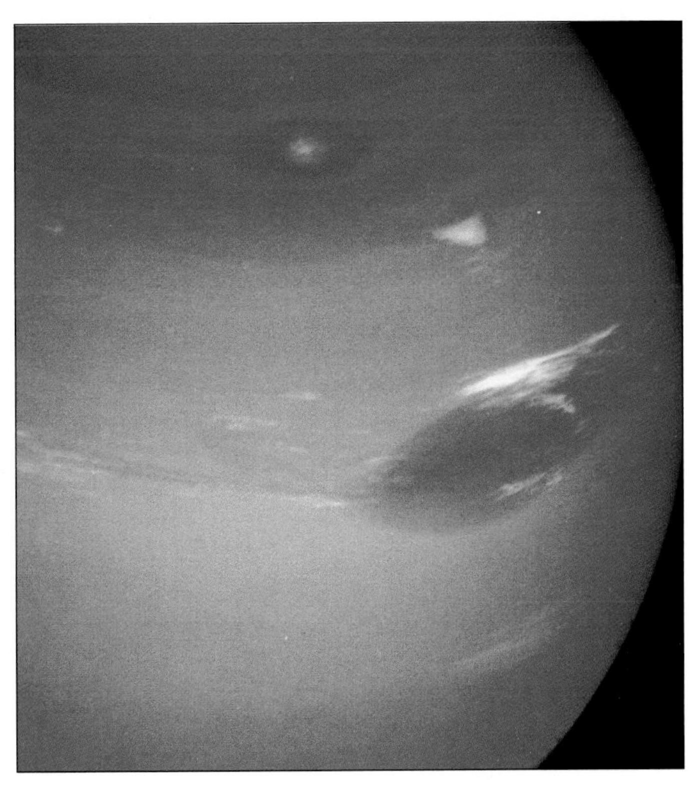

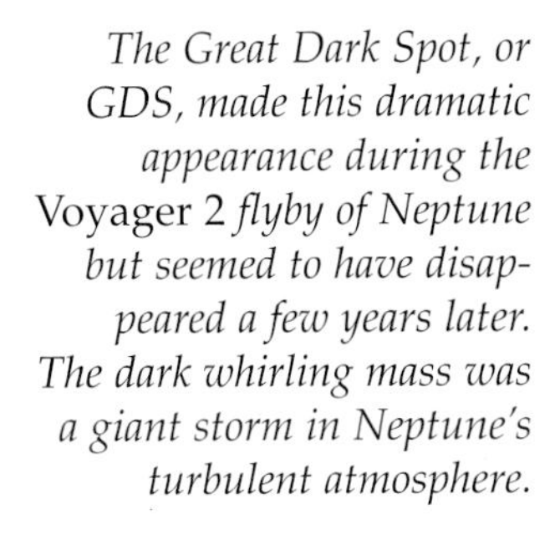

The Great Dark Spot, or GDS, made this dramatic appearance during the Voyager 2 *flyby of Neptune but seemed to have disappeared a few years later. The dark whirling mass was a giant storm in Neptune's turbulent atmosphere.*

Huge, long-lasting storms have been visible at several locations on Neptune. During the *Voyager 2* flyby, one such storm was the largest feature on the planet. Scientists named it the Great Dark Spot (or GDS) because it was noticeably darker than the surrounding clouds. The GDS traveled from east to west across the face of the planet. *Voyager 2* images revealed strands of bright white clouds forming over the edges of the storm. These clouds, which looked something like Earth's cirrus clouds, were higher in Neptune's atmosphere than the GDS itself. Astronomers often compared the GDS with a similar storm on Jupiter, the Great Red Spot. They found, however, that

Scientists nicknamed the large, fast-moving white spot in Neptune's atmosphere Scooter.

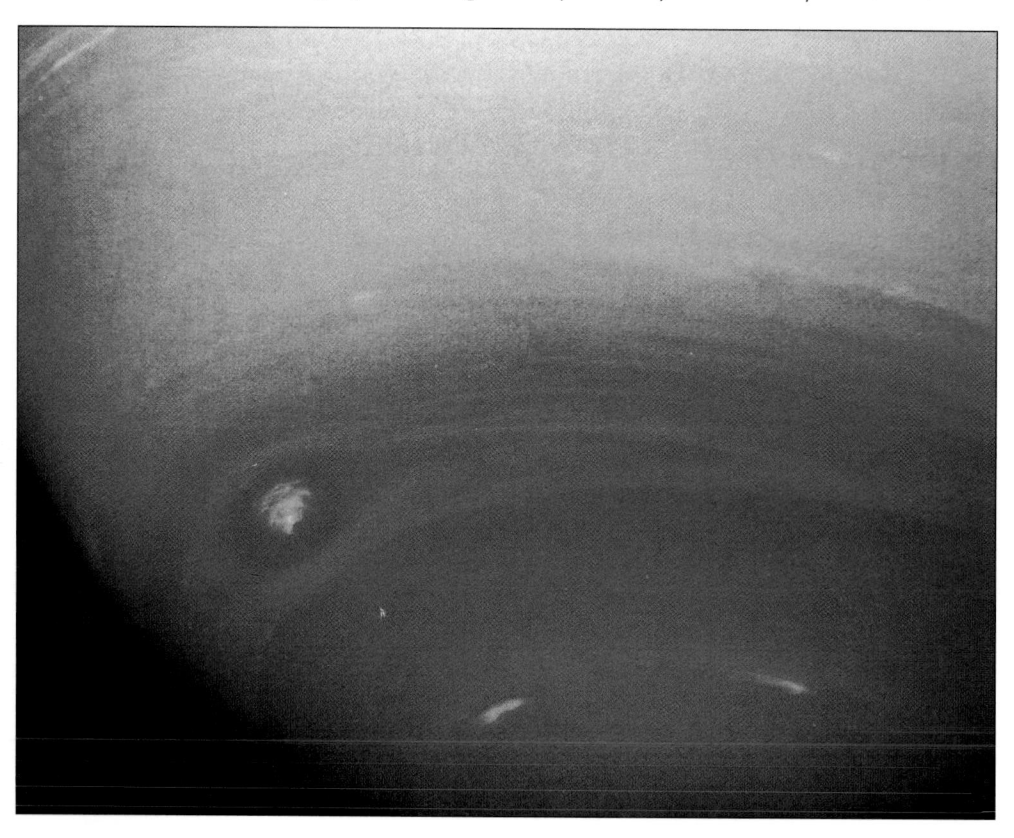

while the Great Red Spot has changed very little over many years of observation, the GDS changed size and shape frequently. Some of the *Voyager 2* photos showed considerable change over a period of just eighteen hours.

In 1994, the Hubble Space Telescope took a series of pictures of Neptune. When space scientists looked at the images, they got a shock—the GDS seemed to have vanished! Either the storm had broken up, or higher clouds were masking the GDS. Meanwhile, a new dark spot, probably another storm, had appeared farther north.

The *Voyager 2* images showed a second dark spot on Neptune. It lay far south of the GDS and traveled from west to east. Scientists called this feature D2. It had a center, or core, of bright clouds, and in some photographs it resembled a dark, almond-shaped eye with a glowing white pupil. Like the GDS, it was a giant storm that was also noticeably absent from the 1994 Hubble Space Telescope photographs.

The third prominent feature that *Voyager 2* detected on Neptune was a large patch of bright cloud that scientists named the Scooter because it traveled around the planet much faster than the GDS. The Scooter was deeper in Neptune's atmosphere than the bright cirrus-type clouds near the GDS. Scientists thought that it might have been a high plume rising from cloud masses in a lower atmospheric level.

The large-scale, rapidly changing storms and cloud features in Neptune's atmosphere suggest that the planet has a very dynamic, lively atmosphere. Its turbulent weather may be caused by temperature differences between the very cold upper layer of the atmosphere and the slightly less cold deeper layers. Neptune's weather may be as changeable and fast-moving as that of Earth, although on a much vaster scale. The GDS, for example, was a single storm about the size of the entire Earth!

4

Moons and Rings

Neptune is not just a planet. It is the center of its own planetary system, just as the Sun is the center of the Solar System. The Neptunian system contains mysterious rings and eight satellites. Six of these satellites were surprises revealed by *Voyager 2*. Another is a large moon that, according to some astronomers, is one of the weirdest and most fascinating worlds in the Solar System.

THE BIG MOON

Triton, the first of Neptune's moons to be discovered, is also the largest—large enough to be seen through Sir William Lassell's nineteenth-century telescope. With a diameter of 1,680 miles (2,705 km), Triton is three-quarters the size of Earth's Moon, but it moves much more quickly. Triton takes 5.877 days to complete a revolution around Neptune. If our Moon revolved that rapidly, a month on Earth would last less than six days.

Space scientists had been hoping that *Voyager 2* would give them a good look at Triton, and they were not disappointed. Neptune's largest moon turned out to be full of surprises. For one thing, Triton is smaller and heavier than the experts expected. It is also extremely cold, with surface temperatures as low as -391 degrees F (-235 degrees C). This makes Triton the coldest planet or moon ever studied by a space probe. Triton is also very bright. It reflects between 60 and 95 percent of the sunlight that hits it. (Compare this with the Earth's

Voyager 2 *passes from the night sky into daylight above Triton, as pictured by artist Julian Baum.*

Moon, which reflects about 11 percent.) Triton's brightness is related to why it is so cold—the satellite reflects so much sunlight that little heat reaches its surface.

Triton's color was another surprise. Much of the moon looks white, but parts of it are pink or reddish. Images from *Voyager 2* showed a bright pink cap covering the south pole. One theory is that the pink regions are covered by nitrogen ice, which could absorb blue light and reflect reddish light. Another possibility is that the reddish

Triton's many-textured surface bears the signs of impacts, past floods, and volcanic eruptions that continue to this day.

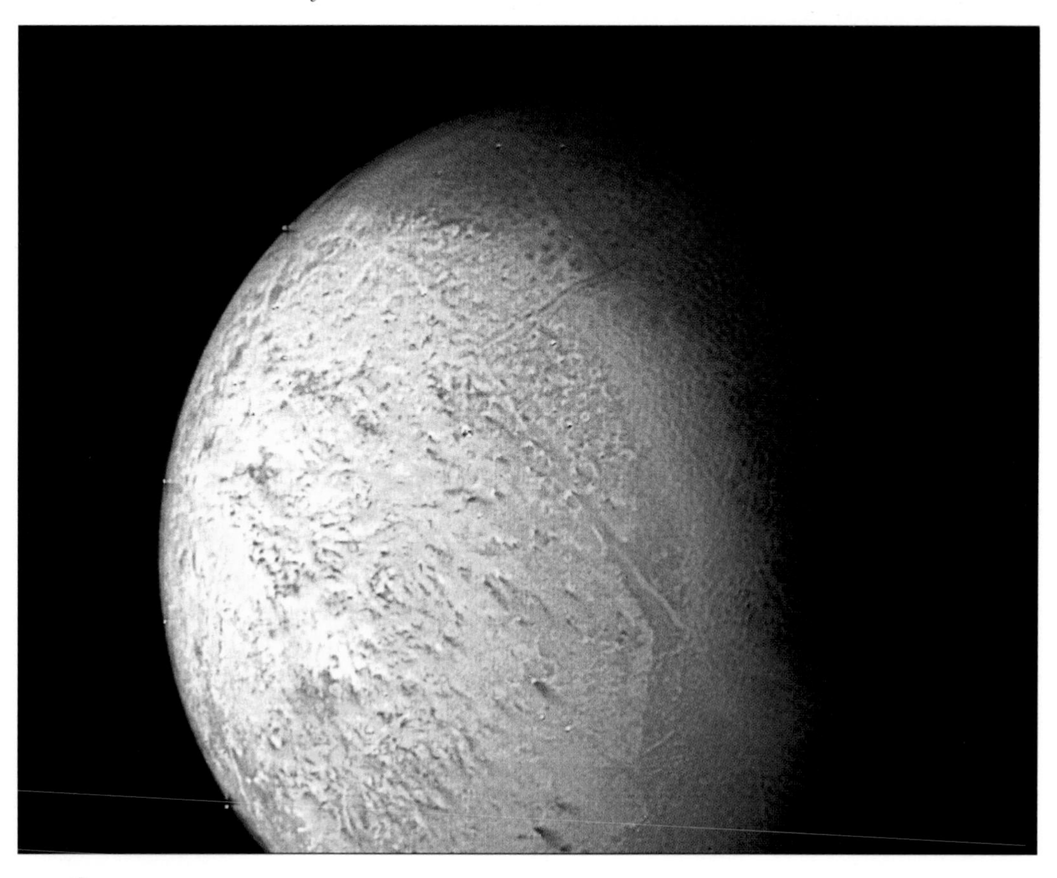

color is caused by chemical reactions set off in the methane snow and ice by cosmic rays and ultraviolet light.

Neptune's largest moon is an icy world, made up of thick layers of frozen methane, frozen nitrogen, and water ice encircling a rocky core. Many of Triton's surface features, such as tall ridges and cliffs, appear to be made of water ice. On Earth such huge structures of ice could not support their own weight, but at temperatures as cold as those on Triton, ordinary water forms ice that is as hard and strong as rock.

Triton's surface is cracked, pitted, wrinkled, and ridged. These markings suggest that the moon has had a complicated history. Undoubtedly, like all of the moons and planets in the Solar System, it has been struck countless times by meteorites and fragments of space debris. Triton has an atmosphere, but it is too thin to offer much protection against meteorites, which in a thicker atmosphere would burn up before hitting the surface. Triton should be covered with impact craters, the way Earth's Moon is. Yet its surface shows no large impact craters and relatively few small ones, leading scientists to believe that Triton's surface crust is less than a billion years old, not several billions like the surfaces of the Moon and most other known satellites. Many of Triton's surface features resemble the remains of great floods or volcanic eruptions. Experts think that a few millions or hundreds of millions of years ago, much of Triton's surface was soft. Internal heat may have warmed it, or perhaps dark matter absorbed enough energy from the Sun to melt the ice. As the icy crust softened, floods filled old impact craters. Slushy ice flowed like lava. As the surface cooled, great cracks and depressions formed, leaving Triton's surface remarkably varied.

Part of the moon's surface is called the "cantaloupe terrain" because it looks a bit like the rind of a cantaloupe melon. This terrain is fairly level but covered with shallow, rounded pockmarks. Long cracks, some of them with ridges at their edges or in their centers,

If you have ever looked at the outside of a cantaloupe melon, you will understand why astronomers dubbed this territory Triton's "cantaloupe terrain." They are still trying to determine how it was formed.

zigzag across this landscape. Elsewhere there are smooth expanses like vast frozen lakes, with terraced walls or shelves around their borders. The terraces may represent the former shores of ice or liquid seas.

Perhaps the most astounding thing about Triton is that it is an active world, not a dead one like our Moon. *Voyagers 2*'s pictures revealed strange dark streaks as much as 60 miles (100 km) long. They are the remnants of material—probably dust, nitrogen gas, and ice—shot out of Triton's interior through vents. *Voyager 2* captured one vent in action, photographing a plume of matter that rose 5 miles (8 km) straight up into the moon's atmosphere and then flattened out to form a 90-mile (145-km) cloud that cast a shadow on the surface below. Scientists have called these eruptions geysers and ice volcanoes. The only thing like them in the Solar System occurs on Io, one of the moons of Jupiter. Triton's volcanic activity was another of *Voyager*'s unexpected findings.

SMALLER SATELLITES

Voyager 2's flight path did not let the probe make a close study of Nereid, the Neptunian satellite discovered in 1949. Nereid has the most eccentric, or oval-shaped, orbit of any known moon. Nereid's distance from Neptune ranges from a minimum of 841,100 miles (1,353,600 km) to a maximum of 5,980,200 miles (9,623,700 km). This highly eccentric orbit resembles the orbits of comets, and some space scientists think that Nereid may be a comet or asteroid captured by Neptune's gravitational pull. Another point in favor of that theory is

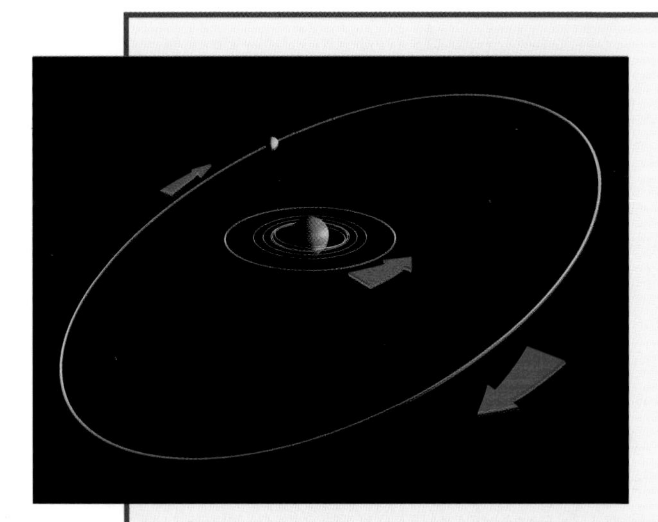

Neptune's inner satellites (and Nereid, not shown in this illustration) revolve around the planet in the same direction as the planet's rotation. Only Triton, shown here as the outer moon, has a retrograde, or backward, revolution, leading some astronomers to suggest that its origin was different from that of the other satellites.

A Backward Moon

Triton is the only large satellite in the Solar System to circle its planet in what scientists call retrograde motion. It revolves around Neptune "backward," in the opposite direction to the planet's rotation. On Earth, both the Sun and the Moon appear to rise in the east and set in the west. On Neptune, the Sun would rise in the east, but Triton would rise in the west. The only other known retrograde satellites—one circling Saturn and four around Jupiter—are very small.

Naiad, Neptune's smallest satellite, is seen here revolving far above the surface of the planet.

that Nereid is much farther from Neptune than any of the planet's other satellites—Triton, the second farthest, is only 219,967 miles (354,800 km) from Neptune. Perhaps Neptune swept up the wandering Nereid, which has a diameter of 211 miles (340 km).

As if to make up for not closely approaching Nereid, *Voyager 2* gave scientists an unexpected bonus: six new satellites. All are smaller than Triton and located within Triton's orbit. Scientists call them the inner satellites. They appear to be rocky, lumpy, irregularly shaped worlds, covered with impact craters. None shows any sign of the kind of geological and volcanic activity revealed on Triton. Although *Voyager 2* discovered these small satellites, it was not able to gather much data on them. Their sizes and the details of their orbits around Neptune are estimates, not precise measurements.

Neptune's smallest moon is also its nearest. Naiad is about 33 miles (54 km) across and revolves around Neptune at a distance of approximately 14,400 miles (23,200 km) near the top of its atmosphere. The second moon, Thalassa, measures 50 miles (80 km) across and is 15,700 miles (25,200 km) from the clouds. Despina, the third satellite, is 90 miles (150 km) across and 17,200 miles (27,700 km) from the planet.

How Moons are Named

When *Voyager 2* sent back pictures of six new satellites orbiting Neptune, space scientists labeled them 1989N1 through 1989N6. Now each of the six has a name drawn from ancient mythology. How did they get their names?

The International Astronomical Union (IAU), made up of astronomers from many nations, is responsible for naming bodies in space and the features on them. Anyone—a scientist, a citizen, even you—can suggest a name for some new object or feature in the Solar System, or ask that a particular name be given to the next thing that is discovered. If the IAU approves the name, it becomes permanent.

Names may be chosen to fit a theme. Both Triton and Nereid, for example, were mythological beings said to follow Neptune, the sea god, and the six newly discovered Neptunian moons also have names from Greek and Roman myths. But scientists who name planetary features now draw on the mythologies of all cultures. Features on Triton's surface bear such names as Abatos Planum (an island sacred to the ancient Eygptians), Hili (a water spirit in the legends of Africa's Zulu people), Kurma (the Hindu god Vishnu in the form of a tortoise), and Ryugu Planitia (the palace of undersea dragons in Japanese legends). All of the place names on Triton are related in some way to water, but they reflect the mythical heritage of the entire Earth.

Galatea is the fourth satellite, about 23,100 miles (27,700 km) above Neptune. Its diameter is around 110 miles (180 km). Larissa is 30,300 miles (48,800 km) above the cloud tops and about 120 miles (190 km) across. The sixth satellite, Proteus, revolves at a distance of approximately 57,700 miles (92,800 km) and measures 250 miles (400 km) in diameter. Proteus is larger than Nereid, but its existence was not confirmed until the *Voyager 2* flyby because, unlike Nereid, it is very dark and difficult to see. All six of the inner moons, in fact, are dark as if they were covered with soot.

NEPTUNIAN RINGS

Some of the early observers who thought they had glimpsed rings around Neptune described them as "ring arcs," sections of rings that did not go all the way around the planet. No one, however, could explain how such arcs or incomplete rings could exist—the laws of physics state that particles revolving around a larger body will spread out to form complete rings around the object. The images from *Voyager 2* settled the matter. They showed that Neptune has complete rings, not partial arcs. The so-called arcs are clumps or concentrations of material in some of the rings.

Unlike Saturn's bright, broad rings, Neptune's are made up of very dark dust that is spread rather thinly (as are Uranus's). The smallest pieces of this ring material are very fine, no larger than fine ash. The rings are extremely hard to photograph and almost impossible to spot from Earth. Still, scientists studying the *Voyager 2* images and working with the Hubble Space Telescope have identified three major rings and several smaller ones. All of them lie between the orbits of Naiad and Proteus. Scientists believe that the gravitational force of Neptune's small inner satellites may affect the distribution of material in the rings, pulling it into thicker clumps in some sections.

The outermost ring is a narrow band 23,700 miles (38,100 km)

Neptune's two major rings are named for John Couch Adams (the outer one) and Urbain-Jean-Joseph Leverrier, whose calculations led to the planet's discovery.

above Neptune's atmosphere. It has been named the Adams Ring. This ring contains three areas of denser and brighter material that are probably responsible for most sightings of "ring arcs." Inside the Adams Ring is the Arago Ring, and inside Arago is a faint band known as the Lassell Ring. The second major ring, named for Leverrier, is 17,700 miles (28,500 km) from the planet's cloud tops. The third major ring, Galle, is a broader and fainter band that begins about 10,600 miles (17,100 km) above Neptune. The Galle Ring may reach all the way down to the planet's atmosphere.

Experts are not completely certain how ring systems formed around the gas giants. Many believe that Neptune's rings were once small inner moons that either collided and smashed to bits or were torn apart by the planet's gravity. New material is continually added to the rings in the form of dust thrown out into space when meteorites strike the remaining moons.

5

MYSTERIES OF NEPTUNE

More than a century and a half after the search for Neptune began, *Voyager 2* and the Hubble Space Telescope gave scientists a wealth of information about Neptune and its planetary system. Some of that information gave rise to new theories—and raised many new questions.

QUESTIONS ABOUT NEPTUNE

Neptune holds many mysteries, but scientists can answer one major concern for certain. Neptune and Pluto will never collide. Even though Pluto crosses Neptune's orbit twice every 248 years, there is no chance that the two planets will ever strike each other like two massive billiard balls.

The orbits of the two planets are stable, and they ensure that whenever Pluto crosses Neptune's orbit, Neptune is far, far away. The two planets never get closer to each other than 17 AU. Just to make sure, experts have created computer programs that model the orbits of the two planets from billions of years ago to billions of years in the

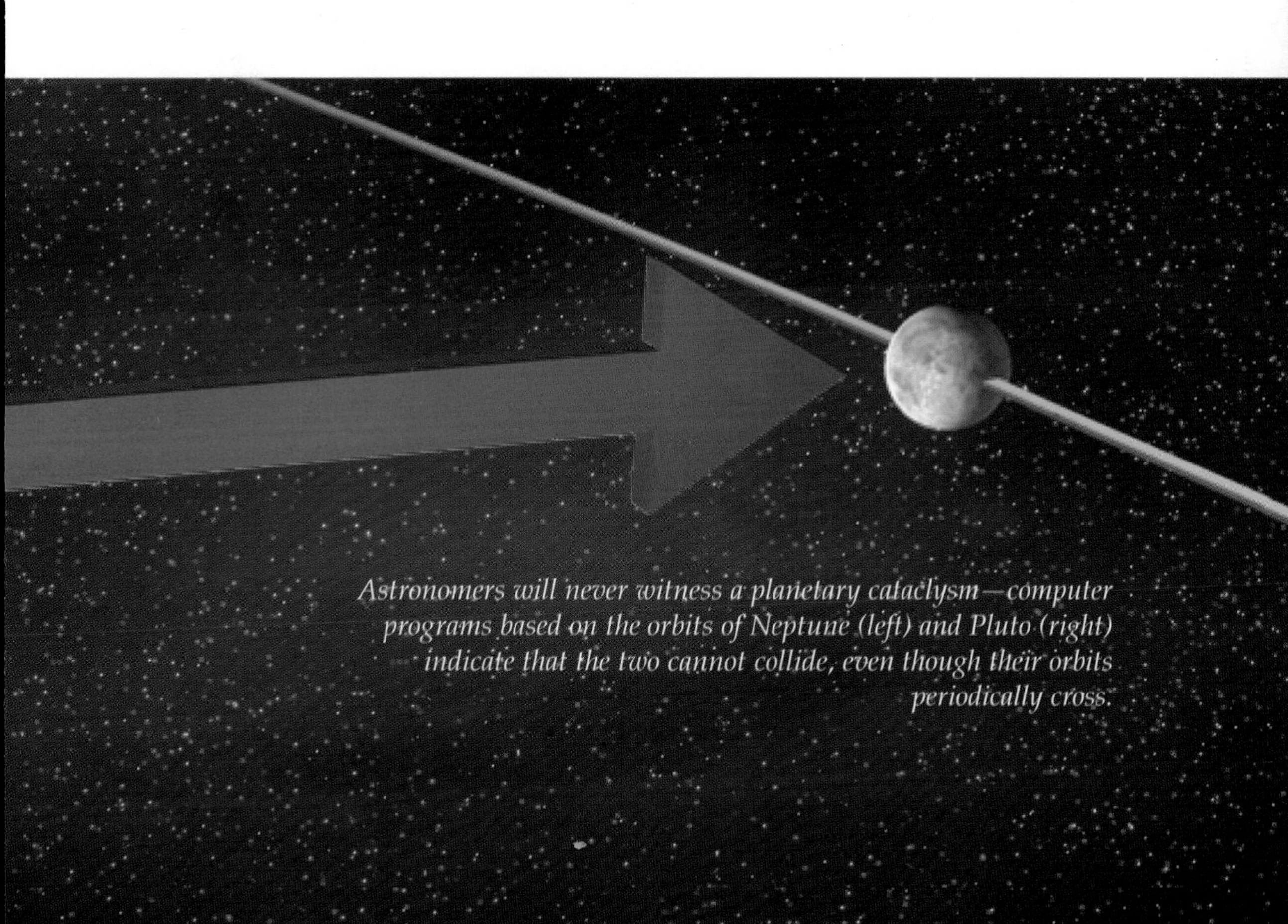

Astronomers will never witness a planetary cataclysm—computer programs based on the orbits of Neptune (left) and Pluto (right) indicate that the two cannot collide, even though their orbits periodically cross.

future. These programs show Neptune and Pluto dancing around and around the Sun but never directly encountering each other.

Neptune plays a key part in an astronomical theory that surfaced in the late 1990s. The theory began with a question: Have the planets always occupied their present positions, circling endlessly around the Sun in the same orbits? Scientists now believe that many of the smaller bodies in the Solar System, such as asteroids, comets, and moons, have moved around quite a lot since the Solar System was formed. Could planets have moved as well?

One clue may lie in the Kuiper Belt, a band of icy asteroids and small planetoids stretching out from Neptune's orbit to at least 9 billion miles (14.5 billion km) from the Sun. Renu Malhorta, a scientist at

Beyond Neptune's orbit lies the Kuiper Belt, a band of countless small bodies circling the Sun. (This illustration shows only half of it.) Some theories about the formation of the Solar System suggest that as Neptune moved outward from the Sun to its present orbit, its gravity drew these asteroids and planetoids into position.

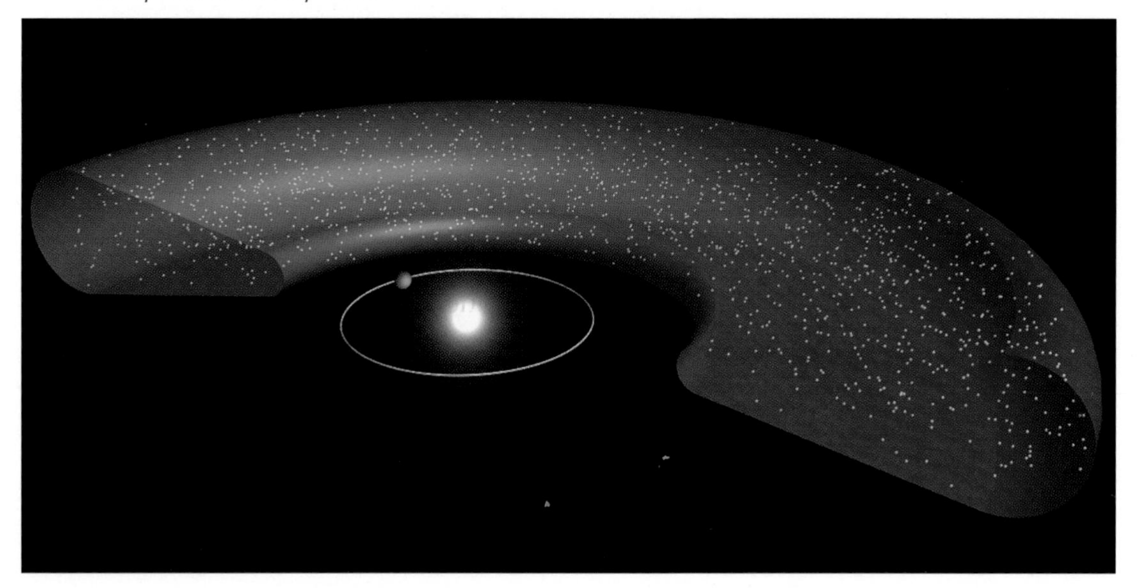

the Lunar and Planetary Institute in Houston, Texas, has studied the way these little-known bodies are distributed in space. She thinks that they may have been shoved into their present positions by slow migrations of the gas giants over millions of years. In particular, she believes that Neptune may have originally formed closer to the Sun, perhaps only two-thirds as far away as it is today. During the early history of the Solar System, Neptune knocked into many smaller orbiting bodies such as planetoids. The energy of these impacts caused Neptune to speed up, and, according to the laws of planetary physics, acceleration made Neptune's orbit grow larger over time. Gradually Neptune moved farther out from the Sun. As this happened, Neptune's gravity pulled many of the remaining planetoids outward as well. Eventually these small bodies collectively formed the Kuiper Belt.

The theory of the migrating planets has not been proven, but astronomers believe that in the coming years more detailed mapping of objects in the Kuiper Belt will either support or overturn the theory. If enough of these Kuiper Belt objects appear to be orbiting in certain mathematical relationships to Neptune's orbit, they may be evidence of an earlier era in the history of the Solar System, when Neptune moved unstoppably outward, pushing and pulling the smaller bodies along with it.

Neptune has other puzzles. Experts in planetary magnetism would like to know more about the dynamic processes in the interior of the planet that have created a magnetic field that is off-center and sharply tilted. Those investigating atmospheric conditions wonder how a planet that receives little energy from the Sun and has a fairly weak internal heat source can produce the Solar System's strongest winds. They would also like to know what happened to the Great Dark Spot and other features photographed during the *Voyager 2* flyby. Other unsolved mysteries concern Neptune's ring and satellite systems, especially the large moon Triton.

QUESTIONS ABOUT TRITON

Triton's retrograde orbit, together with the fact that it is surprisingly heavy, or dense, for its size, make some space scientists think that it did not originate near Neptune. The retrograde orbit means that Triton could not simply have formed where it is from the original cloud of material that was the basis of the Solar System. Such a process of formation would not explain why the satellite revolves backward. But if Triton had already formed and been in motion when Neptune encountered it, Triton may well have then assumed a retrograde motion around the larger planet.

Perhaps Triton started out as a large asteroid, or even a small outer planet like Pluto. At some point after the Solar System formed more than 4.5 billion years ago, Neptune's gravitational pull could have captured Triton and set it into orbit. Such an event would have produced powerful tides deep inside Triton. These tides would have created enough energy to melt the satellite's surface. Scientists at NASA's Jet Propulsion Laboratory have suggested that Triton might have remained in liquid form for as long as a billion years after being captured by Neptune. This would account for Triton's surface features as well as for its continuing icy volcanic activity. Those ice volcanoes are a source of much puzzlement to scientists, who would like to know more about the energy source that fuels them.

Many astronomers believe there is a direct connection between Triton and Pluto. As far as they can tell—and they know very little about Pluto—Neptune's largest moon and the ninth planet are very similar in size, density, and chemical makeup. If they were not small, twin asteroids or planets, possibly both were comets, captured by our Solar System in ages past. Another theory suggests that Pluto, like Triton, was once a satellite of Neptune until some impact—possibly a collision with a third moon that broke up into the six inner satellites—knocked it out of its orbit and sent it reeling to the outer fringe of the system.

would take such a long time, that even the most advanced experts in space travel can barely imagine how it could be done. And there is no urgent reason to do it. As one NASA official said when asked about manned exploration of the distant planets, "That's what probes are good at."

While waiting for the next Neptune probe, planetary astronomers who want to solve some of Neptune's mysteries must rely on two sources of information. One is the vast store of data gathered by *Voyager 2*, which astronomers will continue to study for years, re-examining the data as new theories arise. The other is telescopic observation. Unfortunately, the demand is high for time on the world's largest telescopes and on the Hubble Space Telescope. With some astronomers searching for asteroids that might hit the Earth, others studying the Moon and the nearer planets, and still others probing the depths of space for clues to the origin of the Universe, Neptune is not a high priority. Our knowledge of Neptune has taken an enormous leap forward since 1989, but the planet of cold blue clouds will certainly keep some of its mystery for the rest of our lifetimes.

Neptune Fact Sheet

Mean distance from Sun: 2.797 billion miles (4.497 billion km)
Diameter: 31,404 miles (50,538 km) at the equator
Mean surface temperature: -360 degrees F (-218 degrees C)
Surface gravity: 1.2 times Earth's gravity
Period of revolution (year): 165 Earth years
Period of rotation (day): 16 hours, 7 minutes
Number of satellites: 8 (from nearest to farthest, Naiad, Thalassa, Despina, Galatea, Larissa, Proteus, Triton, Nereid)

Glossary

astronomer one who studies space and the objects in it

atmosphere layer of gases surrounding a world; air

AU astronomical unit, or about 93 million miles (150 million km), the distance from the Earth to the Sun

celestial having to do with the sky, the heavens, or astronomy

gravity force that holds matter together and draws objects toward one another

interstellar having to do with the distances between the stars

orbit path followed by an object as it revolves around another object

probe machine or tool sent to gather information and report it to the sender

satellite object that revolves in orbit around a planet; natural satellites are called moons

sensor instrument that can detect and record information, such as light waves, sounds, X-rays, or gravitational and magnetic readings

Solar System all bodies that revolve around or are influenced by the Sun, including planets, moons, asteroids, and comets

telescope device that uses magnifying lenses, sometimes together with mirrors, to enlarge the image of something viewed through it

Find Out More

Books for Young Readers

Branley, Franklyn. *Neptune: Voyager's Final Target*. New York: HarperCollins, 1992.

Brewer, Duncan. *The Outer Planets: Uranus, Neptune, Pluto*. Tarrytown, NY: Marshall Cavendish, 1993.

Brimner, Larry Dane. *Neptune*. Danbury, CT: Children's Press, 1999.

Landau, Elaine. *Neptune*. New York: Franklin Watts, 1991.

Scott, Elaine. *Close Encounters: Exploring the Universe with the Hubble Space Telescope*. New York: Hyperion Books for Children, 1998.

Simon, Seymour. *Neptune*. New York: Morrow Junior Books, 1991.

Other Books

Burgess, Eric. *Far Encounter: The Neptune System*. New York: Columbia University Press, 1991. Burgess, author of a number of books on the planets and of several NASA project histories, focuses on the *Voyager* 2 mission and explains its origin, goals, progress, technology, and results. Black-and-white drawings and photographs and three dozen images from the *Voyager* flyby of Neptune illustrate the book.

———. *Uranus and Neptune: The Distant Giants*. New York: Columbia University Press, 1988. A summary of scientific knowledge about these two outer planets after *Voyager 2*'s encounter with Uranus but before the Neptune flyby, this volume includes information about Pluto and the theory that the Solar System may include an unknown tenth planet.

Grosser, Morton. *The Discovery of Neptune*. New York: Dover Publications, 1962. This short and highly readable volume tells the story of Neptune's discoverers and the international conflict that arose over the discovery of the planet.

Hunt, Garry, and Patrick Moore. *Atlas of Neptune*. Cambridge, England: Cambridge University Press, 1994. The authors include background on the discovery of Neptune, but their real subject is the 1989 *Voyager 2* flyby and the wealth of new information it yielded about the planet. Triton and the other satellites are also included.

Littmann, Mark. *Planets Beyond: Discovering the Outer Solar System*. New York: John Wiley & Sons, 1988. Much of this book deals with the Uranus and the *Voyager 2* flyby of that planet, but there are two chapters on Neptune. Additional chapters cover the *Voyager 2* Grand Tour, Pluto, and the outer reaches of the Solar System.

Moore, Patrick. *The Planet Neptune*. Chichester, England: Ellis Horwood, 1988. Part of the Library of Space Science and Space Technology series, this book was written before the *Voyager 2* flyby of Neptune in 1989. The author, a distinguished astronomy writer, provides a concise but detailed account of the discovery and study of

Neptune before 1989 and of the questions that scientists hoped *Voyager 2* would answer. Correspondence among the mathematicians and observers who discovered Neptune is included.

WEBSITES

The following Internet sites offer information about and pictures of Neptune, along with links to other sites:

www.bbc.co.uk/planets/ Home page of the British Broadcasting Corporation's Nine Planets site, companion to a television series. One section of the site is devoted to Neptune.

www.vraptor.jpl.nasa.gov/voyager/vgrnep_fs.html *Voyager's* Neptune Science Summary, maintained by NASA's Jet Propulsion Laboratory.

www.pds.jpl.nasa.gov/planets/welcome/html Home page of the Welcome to the Planets site maintained for NASA by the California Institute of Technology.

www.seds.org/nineplanets/neptune/html A Neptune site maintained by the Lunar and Planetary Laboratory of the University of Arizona.

About the Author

Rebecca Stefoff, author of many books on scientific subjects for young readers, has been fascinated with space ever since she spent summer nights lying on her lawn in Indiana and gazing up at the Milky Way. Her first telescope was a gift from parents who encouraged her interest in other worlds and in this one. Today she lives in Portland, Oregon, close to the clear skies and superb stargazing of eastern Oregon's deserts.

Index